Library of
Davidson College

# THE WORLD PRICE OF OIL
## A Medium-Term Analysis

### Hendrik S. Houthakker

American Enterprise Institute for Public Policy Research
Washington, D.C.

Hendrik S. Houthakker is professor of economics at Harvard University and an adjunct scholar of the American Enterprise Institute.

ISBN 0-8447-3224-9

National Energy Study 13, October 1976

Library of Congress Catalog Card No. 76-40596

© 1976 by American Enterprise Institute for Public Policy Research, Washington, D.C. Permission to quote from or to reproduce materials in this publication is granted when due acknowledgment is made.

*Printed in the United States of America*

# CONTENTS

Market Organization     2

A Model of the World Oil Market     3
   Transportation   3
   Refining   5
   The Interdependence of Prices   6
   The Demand for Petroleum Products   6
   The Supply of Crude Oil   11
   North American Production   13

Implementation of the World Petroleum Model     17

Simulation Results with "Optimistic" Elasticities     20

Simulation Results with "Pessimistic" Elasticity Assumptions     23

Constraints on OPEC     25

Possible Countermeasures by Oil-Importing Nations     29

The Outlook for 1985     34

Conclusions     36

Postscript: The Longer Run     36

# THE AEI
# NATIONAL ENERGY PROJECT

The American Enterprise Institute's
National Energy Project was established in early 1974
to examine the broad array of issues
affecting U.S. energy demands and supplies.
The project commissions research into all important
ramifications of the energy problem—economic
and political, domestic and international, private
and public—and presents the results
in studies such as this one.
In addition it sponsors symposia, debates, conferences,
and workshops, some of which are televised.

The project is chaired by Melvin R. Laird,
former congressman, secretary of defense,
and domestic counsellor to the President,
and now senior counsellor of *Reader's Digest*.
The advisory council represents a wide range of
energy-related viewpoints.
Professor Edward J. Mitchell of the
University of Michigan is the project director.

Views expressed are those of the authors
and do not necessarily reflect the views of
either the advisory council and others associated with
the project or of the advisory panels,
staff, officers, and trustees of AEI.

# THE WORLD PRICE OF OIL

To the economist, the problem of energy, sometimes with much exaggeration called "the energy crisis," is primarily a matter of prices. The question facing us is not, and presumably will not be in the future, whether energy supplies will be sufficient to meet the demand, even though that is the formulation often used in popular discussions. Unless price controls are going to be a permanent component of our economic institutions—which I neither hope nor expect—supply and demand in the energy markets are going to balance at some price. The question addressed here is what that price is likely to be. Market equilibrium does not mean that supply and demand are exactly equal, since inventories may fluctuate, but in the energy markets such fluctuations are likely to be important only in the short run. Furthermore, speaking about energy prices means in effect speaking about the price of crude oil. Not only is petroleum the largest component of the world's energy supply, and likely to remain so in the medium term, but its price is central to the determination of most other energy prices. The following discussion will therefore focus on the level and pattern of prices in the world petroleum market. The emphasis will be not only on the medium-term outlook (until the early 1980s) but also (especially in the final section) on what the consuming countries can realistically do about it.

---

Note: Thanks are due to Michael Kennedy and Lawrence Kreicher for some of the calculations underlying this paper, to Data Resources, Inc., for access to its World Petroleum Model, and to Edward Mitchell for valuable suggestions. An earlier version of this paper was presented at an American Enterprise Institute conference in early October 1974; for the discussion there, see Edward J. Mitchell, ed., *Dialogue on World Oil* (Washington, D.C.: American Enterprise Institute, 1974).

## Market Organization

Some understanding of the organization of the world petroleum market is useful to this analysis. Unlike prices of many other raw materials, the world price of oil has not been determined in any central location; at least it was not until 1973, when the Organization of Petroleum Exporting Countries (OPEC) took charge. For oil, there has been nothing like the Chicago wheat market or the London copper market, where prices are set daily by the offers of producers and the bids of consumers, with considerable participation by merchants and speculators.

Why does petroleum lack such a central market? In common with most commodities that are traded on futures markets, petroleum is storable. While not as homogeneous as copper, it is not more heterogeneous than wheat, and a serviceable standard contract would not be hard to design. Although transportation costs are relatively more important than for most centrally traded commodities, this would not seem an insuperable obstacle either. Perhaps the main reason for the failure of a central market to develop is that for many years the industry has been dominated by integrated companies that handle oil from the well to the gasoline pump. Merchants, brokers, and other intermediaries are relatively unimportant; as a result, arm's-length transactions have traditionally been less prevalent in petroleum than in many other raw materials.[1] The significance of this point is that the integrated companies appear to be losing much of their control over crude oil, so that arm's-length transactions will become more common. In due course, a central market may emerge.

Whatever the reasons for the absence of a central mechanism, it has not prevented the petroleum market from being truly worldwide. Of the major producing and consuming countries, only the Soviet Union and China have stood largely aside. The Soviet Union has not been a major factor in the world petroleum market in recent years;

---

[1] The copper industry also has a number of partially integrated major companies, especially in the United States, but these companies are at the same time large sellers of refined copper to independent fabricators. Outside of the United States, there is not much vertical integration in copper.

To avoid misunderstanding I should add that the preceding remarks are not intended to support the notion of breaking up the major oil companies. Large though some of them are, they operate in a large and risky industry and I am not aware of evidence they are *too* large. Nevertheless, competition in the petroleum industry is probably not as vigorous as it should and could be. Apart from government intervention (for instance, through price controls), a possible cause of this lack of vigor appears to be widespread use of joint ventures, which is now being curtailed.

together with its Eastern European clients it is roughly self-sufficient. The Soviet Union may well become a substantial net exporter because of recent large discoveries of oil and natural gas. However, domestic consumption is also growing, and within the five or ten years covered here, Russian exports to the outside world probably will not increase enough to affect the world price of oil. Much the same holds for Chinese exports.

## A Model of the World Oil Market

In analyzing supply and demand in the world market, one must recognize that "supply" in this context means the supply of crude oil (including condensate and natural-gas liquids), since petroleum products by and large cannot be economically produced from other raw materials.[2] Similarly, "demand" means the demand for petroleum products, there being no significant other uses for crude oil.[3] In other words, the demand for crude oil is entirely a derived demand, and the supply of petroleum products is entirely a derived supply.

In addition to primary supply and ultimate demand a description of the world petroleum market must cover the transformation of crude oil at the wellhead into petroleum products delivered to consumers. This transformation has two aspects, refining and transportation. Most of the world's refining industry is located near consuming centers (although there are some significant exceptions such as the Caribbean refineries) because refining involves very little lost weight and because it generally costs more per barrel to transport products than crude. This is an important point, because a number of oil-producing countries have announced plans for the establishment of export refineries that would require subsidies in one form or another to make them viable.

**Transportation.** The bulk of international transport in the world petroleum market is by tanker rather than pipeline. Entry by independents is sufficiently easy to make the tanker market highly competitive, even though the major oil companies own many tankers and charter additional tonnage on long-term contracts. The tanker industry has recently seen a considerable increase in productivity resulting from the introduction of ever-larger vessels. Although some tankers are

---

[2] The probable exceptions are oil shale and tar sands, the production of which is likely to be of minor importance until the 1980s.
[3] Some crude oil is burned by electric-power stations, but usually only after the more volatile fractions have been removed by "topping."

now so large that, because of draught and length limitations, they can operate only on very few routes, considerable scope remains for the replacement of small tankers by larger ones. The lower cost of large tankers means that regions where they cannot be used (such as the United States, whose ports are generally not deep enough for tankers over 80,000 tons) are at a disadvantage in foreign trade.

In the short run, tanker rates fluctuate sharply; spot rates, usually expressed in terms of "world scale," can fall quite rapidly—say, after a shipbuilding boom; but a shift in transportation patterns—for instance, because of the closing of the Suez Canal—can turn them around sharply. Just in the last few years world scale has been as low as 25 and as high as 300.[4] Such a wide price range is found in few other markets; it exists largely because many tankers are operated by their owners or under long-term charters so that the spot market is relatively thin. However, the rates at which long-term contracts are fixed usually are influenced by the prevailing spot rate.

Projections of the world petroleum market in the medium or longer run do not require much attention to spot tanker rates. Because of the relative ease of entry it may be assumed that spot rates fluctuate around a level just high enough to cover operating expenses plus a normal return on the capital invested, possibly with some allowance for the high price risks inherent in the industry. New building, laying-up, and scrappage constitute the mechanism by which this average level is maintained in the medium run. It will be assumed, therefore, that the present large surplus of tankers is eliminated by 1980.

On any particular route between two ports the differential between the price of a specified type of crude oil at origin and at destination (disregarding import and export duties) cannot stay above the freight rate for any length of time if there is reasonably free competition. In the medium or long run, therefore, this differential cannot exceed the average level of the freight rate on that particular route, which in turn is determined by the largest tanker capable of operating between the two ports. Conversely, no oil will flow from one port to another unless the price differential is at least equal to the freight rate. For petroleum products the same propositions hold, but for most of them the relevant freight rate must be increased by the premium charged for "clean" tankers (those used for refined products other than residual fuel oil).

---

[4] When world scale, which is revised annually, equals 100, a hypothetical tanker (rather smaller than the present average) can just cover its total cost. Large tankers can break even at much lower rates. On tankers and many other matters, M. A. Adelman, *The World Petroleum Market* (Baltimore: The Johns Hopkins University Press, 1972) is invaluable though somewhat overtaken by events.

The preceding argument implies a special role for a region that exports to all or most of the other regions. The export price in that region serves as a base from which the import prices elsewhere can be calculated. In the absence of such a central region the price pattern is more complicated but can still be calculated from the principles just stated.

**Refining.** With considerable qualifications this competitive model can also be applied to the refining sector of the world petroleum market. Entry into refining is not as easy as entry into the tanker business because of the absence of central markets in which everyone can obtain crude or dispose of products. The major oil companies refine most of the crude they produce in their own plants and sell it—often under brand names—to ultimate consumers, although a certain amount of trading (sometimes in the form of barter) takes place among them. Nevertheless, the refining sector is by no means completely closed: there are independent crude producers who sell to independent refiners on long-term contracts and other arrangements, and products can also be marketed without relying on the majors, though sometimes only at a discount. In the middle or long run, therefore, a substantial degree of competition may be assumed in the refining sector, so that the capital invested in this sector must earn an approximately normal return.

Just as there are many possible routes to be considered in the shipping sector, so there are many production processes to be considered in the refining sector. In most of these processes certain equipment is used to transform crude of given specifications into certain quantities of final or intermediate products, usually with such additional current inputs as labor or purchased electricity. Sometimes the raw material used in the process is an intermediate product such as naphtha. To calculate the net return to the capital invested in a process, the value of the inputs (including refining costs other than raw materials) has to be deducted from the value of the output. Under competitive conditions the net return to capital in any refining process cannot (at least in the long run) exceed the normal return to capital, nor will any process be used unless it yields a return approximately equal to the normal. Given these propositions and a knowledge of the technological coefficients, it is possible to infer the value of crude from the value of the products or vice versa.[5]

---

[5] For this purpose, the refining technologies need not be the same all over the world, though in practice they are not likely to differ much. The pecuniary costs of refining processes (labor, power, capital charges) may also vary geographically. For a still more accurate description of the refining sector, economies and diseconomies of scale should be taken into account.

**The Interdependence of Prices.** In the two sectors of the world petroleum market discussed so far—transportation and refining—the prices of crude oil and products in different locations are related to each other; these interrelations are now the subject of disagreement among the members of OPEC. Given the price of crude in one location that supplies the rest of the world, one can, in principle, calculate the entire world price structure not only of crude but also of products. Of course, not all the world's crude comes from one point; nevertheless, the Persian Gulf occupies so central a place in the world market that the price there determines all other prices either directly or indirectly, at least in the absence of price controls or quantitative restrictions on trade flows (such as the now-defunct U.S. quotas). The only exceptions would be regions that neither import crude from the Persian Gulf nor export crude or products to third regions where there is such competition. At present there do not appear to be any such regions outside of the Communist area, which will not be considered in this paper.

The price of Venezuelan crude, for instance, follows from the fact that it competes in Europe and North America with Persian Gulf crude, so that the landed price of the two must be approximately equal. The price of the latter is given by adding appropriate freight charges (plus import duties, if any) to the export price in the Persian Gulf, and the export price in Venezuela is then established by subtracting the freight from there to Europe or North America.

Adjustments for the quality of crude (especially gravity and sulphur content) can be made by considering the relative costs of competing refinery processes that use different types of crude to obtain the same products. Thus, it generally costs less to produce gasoline from light crudes than from heavy crudes, and since gasoline is one of the more valuable products, a barrel of light crude is usually worth more than a barrel of heavy crude at the same location. The ratio of the values of these two types of crude, however, does not have to be identical everywhere; in a region in which nearly all the demand is for the heavier products (such as residual fuel oil) the lighter crude may be worth less. Thus, the lighter crudes tend to move to those areas whose demand for light products is relatively strong.

The calculation of the relative prices of different crudes and products at different locations for given demand is a straightforward exercise in linear programming. When demand changes, the differentials generally change too, and one of the difficulties facing a cartel is to adjust them so that members will not underbid each other.

**The Demand for Petroleum Products.** The other two sectors of the world petroleum market—crude supply and product demand—require a somewhat different treatment because the magnitudes in these sectors are themselves dependent on prices. The common assumption that demand for petroleum products in some area is fixed or grows at some fixed rate is not only incorrect in theory, but also highly misleading in practice. The justification for using it in projections of the world market is that energy is "essential" and therefore will be bought at any price.

This argument will not bear examination. Almost without exception, energy is used not for its own sake but as an input into a process from which a desired good or service is obtained. Thus, gasoline is used as an input into many transportation processes (which are themselves generally of an intermediate nature). It follows that such energy cannot be more "essential"—whatever that may mean—than the objects being transported. The energy used to drive people to a cocktail party is at least as dispensable as the party itself—indeed even more so, since there are ways of saving energy in producing the latter: the party-goers may walk rather than ride.

That an increase in a country's aggregate output is generally accompanied by an increase in its energy consumption does not mean that energy is "essential." Any increase in aggregate output is typically accompanied by increases in all sorts of consumption (food, entertainment, paper, tourism, and so on), but to say that all these are "essential" is an abuse of language. For much the same reasons the assertion that energy is (or was) "cheap" in some absolute sense must be dismissed as meaningless.

A realistic treatment of the demand for petroleum products (which serve mostly as sources of energy) must consider them in the same fashion as any other commodity. The aggregate demand for any commodity normally depends on three factors: national income, the price of the commodity itself, and the prices of other commodities, especially those that can serve the same purpose (substitutes) or that are used in conjunction with it (complements). Sometimes additional factors should be taken into account; thus, the demand for heating fuels obviously depends on temperature.

The empirical analysis of energy demand presents considerable technical difficulties. Although time series on consumption, prices, and incomes are available for several countries on an aggregate basis, they generally do not provide enough information for useful estimates of price and income elasticities. Recent studies of energy demand have therefore relied on data for smaller areas within a country, such as the states of the United States,[6] or on combinations of time series for

---

[6] For instance, see H. S. Houthakker, P. K. Verleger, and D. P. Sheehan, "Dynamic Demand Analyses for Gasoline and Residential Electricity," *American Journal of Agricultural Economics*, vol. 56 (May 1974), pp. 412-18.

several countries.[7] The results of these analyses differ, and further work is needed to achieve a consensus. Nevertheless, the following conclusions can already be drawn with some confidence:

(1) Price and income changes have a highly significant effect on the demand for most petroleum products and for electricity (some of which is produced from petroleum).

(2) The effect of price and income changes is greater in the long run than in the short run, principally because energy products are used only in conjunction with capital goods such as cars and electrical equipment. A full adjustment of energy consumption to price and income changes requires time-consuming alterations in the stock of energy-using capital.[8]

(3) For most energy products, the price elasticity is smaller (in absolute value) than the income elasticity. This means that energy consumption in the next several years will be affected not only by the continuing adjustment to higher prices but also by the growth of real output in the principal energy-consuming countries. To some extent this growth rate will itself be affected by the uncertainties and balance-of-payments problems associated with recent developments in the world petroleum market.

(4) Income and prices, and occasionally temperature, appear to explain most of the past behavior of energy consumption. In particular, there is little evidence of general "trends" that would push up energy use regardless of price and income developments. The many projections that rely only on such trends (usually identified with past growth rates) should be viewed with the greatest suspicion.

Now that data for 1975 are becoming available, it is possible to look at the initial effect of much higher prices on consumption. Table 1 presents consumption data for five important countries; it shows that the worldwide boom of 1972–74 was accompanied first by a strong increase in oil consumption, but that the trend reversed itself in 1974. In fact, total consumption in the five countries was significantly lower

---

[7] See M. Kennedy, "An Economic Model of the World Oil Market," *Bell Journal of Economics and Management Science*, vol. 5 (Fall 1974), pp. 540-77; and William D. Nordhaus, "The Demand for Energy," Cowles Foundation Discussion Paper, Yale University, 1975, processed.

[8] Thus, the recent sharp increase in gasoline prices all over the world can be expected to lead ultimately to a lower stock of cars than otherwise would have emerged. Most of these cars will also use less gasoline than cars did before the price increase. For a recent survey of demand elasticities, see the article by K. W. Hoffmann and D. Wood in *Annual Review of Energy, 1976*.

## Table 1
## INLAND OIL CONSUMPTION IN SELECTED COUNTRIES, 1972–75
(amounts in millions of barrels per day)

| Country | 1972 Amount | Percent Change to 1973 | 1973 Amount | Percent Change to 1974 | 1974 Amount | Percent Change to 1975 | 1975 Actual |
|---|---|---|---|---|---|---|---|
| United States | 16.4 | + 6 | 17.3 | − 4 | 16.6 | − 2 | 16.3 |
| Japan | 3.6 | +14 | 4.1 | − 3 | 4.0 | − 8 | 3.7 |
| France | 2.0 | +12 | 2.2 | − 6 | 2.1 | −10 | 1.9 |
| United Kingdom | 2.0 | + 1 | 2.0 | − 6 | 1.9 | −12 | 1.6 |
| West Germany | 2.5 | + 7 | 2.7 | −11 | 2.4 | − 5 | 2.3 |
| Total | 26.5 | + 7 | 28.3 | − 5 | 27.0 | − 5 | 25.8 |

**Note:** Percentage changes calculated from unrounded data.

**Source:** Central Intelligence Agency, *International Oil Developments*, February 12, 1976.

in 1975 than in 1972. Some of the decrease from 1974 to 1975 can be attributed to the worldwide recession of those years, but real GNP declined from 1974 to 1975 by only 2 or 3 percent in four of the countries while in Japan it actually increased slightly. Preliminary figures for the first half of 1976 indicate a revival of consumption resulting from the economic recovery, but the level is still generally below that of 1973.

Higher prices almost certainly were the main cause of the declines in oil consumption. Another factor may have been the availability of natural gas: in the United States gas supplies became increasingly inadequate, forcing some industrial consumers to turn to oil or electricity (and thus indirectly to oil or coal); in Western Europe, on the contrary, natural gas became much more widely available as new pipelines were built following discoveries in the Netherlands, the North Sea, and elsewhere. Consumption in most of the countries was also discouraged by increases in excise taxes (on top of the rise in pre-tax prices), but not in the United States, where price controls prevented a full pass-through and excise taxes remained unchanged. Finally, the winters of 1974–75 were generally mild.

It is too early, however, to determine the relative importance of these various determinants of demand in recent years. With a view to the uncertainty about demand elasticities, different assumptions (optimistic and pessimistic) are made in the projections reported later in this paper.

So far the discussion of the demand side has concentrated on current consumption. In the short run, demand has another important component, the change in inventories. Although of less significance in the longer run, the size, location, and composition of inventories can have a major impact on the price level in the short run. Pre-embargo data on oil stocks on a consistent international basis are hard to come by, but it appears that the increase in the Persian Gulf price of crude at the end of 1973 would not have been as steep if inventories in consuming countries had been more nearly adequate.[9] As in many other raw materials, the worldwide economic boom of 1972–73 may have depleted oil inventories to a level that made importers unusually dependent on current arrivals, hence the panic that seized a number of industrial countries after the embargo of October 1973. This panic (the word is not too strong) in turn demonstrated to the exporters how easy it would be to enforce a much higher price.

---

[9] In the United States, oil stocks during the first half of 1973 were well below the levels of 1970-72 despite considerable growth in consumption. The American Petroleum Institute statistics from which this statement is derived are somewhat different from those used in Table 2.

Since then the inventory situation has improved somewhat but not a great deal. In the United States, for instance, stocks of crude and products rose by about 8 percent between September 1973 (just before the embargo) and one year later, but they fell again in the following year. Much the same is true for other important consuming countries, as shown in Table 2. Thus, the buildup of inventories by consumers, a natural and appropriate reaction to the events of late 1973, turned out to be short-lived. The inventory situation looks somewhat better on a basis of days of consumption, also shown in Table 2. Even though fewer barrels were in storage on September 30, 1975, than a year earlier, they represented about the same number of days because demand had fallen. After that date the inventory picture has not changed a great deal.

The major consuming countries are therefore only marginally less vulnerable to a supply interruption than they were in 1973; moreover, their bargaining power, to the extent that it depends on inventories, has not increased much. Below, I shall consider the prospects for oil prices under alternative assumptions concerning supply and demand. Since this paper is concerned primarily with the medium term, there is no need to pursue the analysis of inventories on a more conceptual level, but plainly they provide little insurance against adverse developments. The United States has now embarked on a stockpiling program which will take several years to reach useful levels.

**The Supply of Crude Oil.** The fourth sector of the world petroleum market, crude supply, presents more difficult analytical problems than do the other three sectors. Transportation and refining can be dealt with by linear programming, and product demand by appropriate extensions of standard techniques of consumption research; moreover, these approaches do not seem to be seriously in error. In the present market structure, however, the analysis of crude supply calls for less familiar tools. While valuable theoretical and empirical work on this subject has been done by Fisher, Erickson and Spann, Epple, and others, it has dealt with the relatively competitive context of the U.S. and Canadian crude markets.[10] The outstanding fact in today's world petroleum market is the emergence of the strong producer cartel, OPEC. Economists do not have much solid knowledge concerning the operation of such a cartel

---

[10] See F. M. Fisher, *Supply and Costs in the U.S. Petroleum Industry* (Baltimore: The Johns Hopkins University Press, 1964); E. W. Erickson and R. M. Spann, "Supply Response in a Regulated Market: The Case of Natural Gas," *Bell Journal of Economics and Management Science*, vol. 2 (Spring 1971), pp. 94-121; and D. N. Epple, *Petroleum Discoveries and Government Policy* (Cambridge, Mass.: Ballinger, 1975).

## Table 2
### OIL STOCKS IN SELECTED COUNTRIES
(amounts in millions of barrels)

| Country | September 30, 1973 | | September 30, 1974 | | September 30, 1975 | |
|---|---|---|---|---|---|---|
| | Amount | Days of consumption | Amount | Days of consumption | Amount | Days of consumption |
| United States | 893 | 52 | 962 | 58 | 944 | 57 |
| Japan | 306 | 75 | 365 | 91 | 340 | 92 |
| France | 194 | 88 | 242 | 115 | 224 | 118 |
| United Kingdom | 153 | 77 | 175 | 94 | 153 | 94 |
| West Germany | 172 | 64 | 177 | 74 | 184 | 80 |
| Total | 1,718 | 61 | 1,921 | 71 | 1,845 | 72 |

**Source:** For amounts, same as Table 1. Days of consumption calculated by dividing amounts of stocks by consumption in Table 1.

and the reactions it may elicit among consuming nations.[11] To sidestep this difficulty it will be assumed that OPEC will operate as a unit (that is, without significant internal problems) and that consumers will continue to act as price takers (that is, without attempting concerted counteractions). These simplifying assumptions will be reconsidered subsequently.

More specifically, it will be assumed until further notice that the price of crude in the Middle East and Africa is fixed at some level that itself is considered a parameter, and that the quantities of crude available at that price are always sufficient to meet the demand. As pointed out above, the Persian Gulf price is the kingpin of the entire oil price pattern in the non-Communist world. In conjunction with the parameters describing the three other sectors of the world market, this price (which is assumed to apply throughout the Middle East and Africa) determines the crude prices charged by OPEC members outside of that area. In any case, these outside members (principally Venezuela and Indonesia) are not big enough to challenge the price-making powers of the Persian Gulf producers in the foreseeable future. The assumption that OPEC will not break down therefore means more specifically that the Middle East and African producers will maintain a common front. It also means that the supply from outside members cannot increase by very much in response to higher prices.

**North American Production.** For the non-OPEC producers the situation is different. At the moment these are primarily to be found in North America, though Europe is rapidly becoming a significant factor (about which more below). The studies noted above suggest that the North American supply is quite responsive to price changes once sufficient time for adjustment has elapsed. This response takes two forms: more exploratory drilling and more production from known reservoirs. Both require time, but the former naturally requires longer. The price for "new" crude in the United States has tripled since late 1973 and the downward trend of exploration has already sharply reversed. Efforts to obtain more oil from old fields (of which the United States has more than the rest of the world) are also proceeding, though these may be held back to some extent by the much lower price of "old" oil under existing price controls. The papers cited suggest that the supply elas-

---

[11] For a survey of recent work on OPEC, see D. Fischer, D. Gately, and J. F. Kyle, "The Prospects for OPEC: A Critical Survey of Models of the World Oil Market," *Journal of Development Economics*, vol. 2 (December 1975), pp. 363-86. This paper, and most of the papers discussed there, appeared too late to be taken fully into account here.

ticity in the United States may be unity or even higher, but this figure is unduly optimistic in present circumstances.

The fact is that the recent development of crude-oil output in the United States does not suggest a supply elasticity anywhere near unity, or indeed exceeding zero. As Table 3 shows, output grew at varying rates in the 1950s and 1960s, and rather strongly through the late 1960s and 1970. But then the trend reversed, with especially steep declines in 1974 and 1975 despite much higher real prices for crude. Until 1974 drilling activity also declined fairly steadily, and an even sharper decline occurred in new oil wells, attributable in part to an increasing emphasis on natural gas. The upsurge in exploratory drilling, and even more in development drilling, that started in 1974 was particularly marked in oil, however.

What Table 3 suggests is that the fall in crude output during the first half of the 1970s resulted from the low level of drilling through 1973, which was itself the consequence of a decline in the relative price of crude oil from 104.3 in 1960–64 to 96.1 in 1970. Not enough new wells were drilled to overcome the natural decline in existing reservoirs. More recently, higher relative prices have encouraged drilling, which should in due course (perhaps as early as 1977 or 1978, assuming price controls are phased out on schedule) lead to a turnaround in domestic crude output in the forty-eight contiguous states— the so-called lower 48. If this analysis is correct, the recent decline in oil production cannot be interpreted as evidence of a negligible supply elasticity, and merely confirms previous findings of long lags in supply response.[12]

Much of the pessimism concerning U.S. production is based on estimates of "proved reserves," which reached a peak of 31.8 billion barrels at the beginning of 1962 for the lower 48 and declined to 24.6

---

[12] Table 3 does not support the widespread belief that exploration in the United States is increasingly unsuccessful because all the good prospects have already been found. The combined success ratio for oil and gas wells did decline through 1971, but since then it has gone up strongly and in 1974 and 1975 it was higher than in any year after World War II. Admittedly, this ratio tells little about the size of discoveries (on which there are only fragmentary statistics so far); no doubt the higher prices of 1974 and 1975 led to the acceptance as commercial of certain wells that would have been abandoned at earlier prices. Nor does it appear that the recent increase in the success ratio is due to deeper (and hence costlier) drilling, since according to preliminary figures the average depth per well was less in 1974 and 1975 than in prior years. Perhaps technological improvements have promoted greater accuracy in exploration. Offsetting these favorable developments, however, is increasing environmentalist opposition to offshore drilling, as a result of which some promising discoveries in California have remained in limbo and leasing along other coasts has been delayed.

## Table 3
### CRUDE OIL: PRICE, OUTPUT, AND DRILLING IN THE UNITED STATES

| Years | Relative Price[a] | Output | | Exploratory Wells | | | | Development Wells | |
|---|---|---|---|---|---|---|---|---|---|
| | | Amount | Percent change | All | Oil | Gas | Success ratio[b] | Oil | Gas |
| 1950–54 | 99.2 | 6,124 | +4.7[c] | 12,180 | 1,818 | 574 | 19.6 | 23,559 | 2,976 |
| 1955–59 | 106.1 | 6,978 | +2.2[c] | 14,451 | 1,979 | 859 | 19.6 | 26,158 | 3,662 |
| 1960–64 | 104.3 | 7,341 | +1.5[c] | 10,981 | 1,244 | 739 | 18.1 | 19,644 | 4,537 |
| 1965–69 | 99.6 | 8,649 | +3.9[c] | 9,447 | 1,033 | 569 | 17.0 | 14,880 | 3,490 |
| 1970 | 96.1 | 9,637 | +4.3 | 7,693 | 790 | 481 | 16.5 | 12,230 | 3,359 |
| 1971 | 99.4 | 9,463 | −1.8 | 6,922 | 651 | 437 | 15.7 | 11,207 | 3,393 |
| 1972 | 95.5 | 9,441 | −0.3 | 7,539 | 684 | 601 | 17.0 | 10,622 | 4,327 |
| 1973 | 93.5 | 9,208 | −2.5 | 7,466 | 619 | 900 | 20.3 | 9,283 | 5,485 |
| 1974 | 132.3 | 8,765 | −4.8 | 8,619 | 814 | 1,195 | 23.3 | 11,970 | 6,045 |
| 1975 | 140.5 | 8,351 | −4.7 | 9,258 | 988 | 1,171 | 23.3 | 15,350 | 6,331 |

Units: Output (including lease condensate but excluding natural gas liquids) in 1,000 barrels per day.

[a] Wholesale price index for crude petroleum divided by overall WPI.
[b] Total number of oil and gas wells as percentage of total number of exploratory wells.
[c] Annual growth rate for five years through last year of period.

**Sources:** American Petroleum Institute, *Basic Petroleum Data Book* (1975); *Oil & Gas Journal* (for 1975); U.S. Bureau of Labor Statistics (for price indexes).

billion barrels at the end of 1974[13] (these estimates disregard the 9.6 billion barrels in Prudhoe Bay, Alaska). The main problem with these widely accepted numbers is that they ignore the dependence of recoverability on price. The following quotation puts them in perspective: "A very large untapped domestic petroleum resource, which could be as large as 100 billion barrels, lies in partly depleted oil reservoirs, and reservoirs no longer capable of production using present technology."[14] Although the fraction of these 100 billion barrels that could be recovered at present prices is not known, the usually quoted reserve figures must be seriously biased downward since they do not include any already discovered oil that could not be recovered at pre-1974 prices but could be recovered at present prices. In fact, the official reserve figures deserve little credence until they have been adjusted to reflect the drastic change in prices during the last three years. The recent sharp increase in development drilling, largely in old reservoirs, suggests that the quotation just given is consistent with industry thinking.

It would therefore be premature to write off domestic oil production as a lost cause, even in the face of widespread skepticism about a significant expansion of U.S. oil output including Alaska.[15] It now appears, in fact, that U.S. production, excluding natural gas liquids, bottomed out at slightly above 8 million barrels per day (mmb/d) in early 1976. In the projections that follow, two sets of assumptions are therefore made about long-term supply elasticities: one optimistic, with an elasticity for the United States of 0.5 (still well below the value implied by earlier research); and one pessimistic, at a value of 0.15. These same elasticities are applied to Canada, where the petroleum situation is similar to that of the United States except that Canada now levies an export duty that keeps the price below the world level. This levy is assumed to have been phased out by 1980.

Apart from price elasticity, two factors determine future crude supplies in North America. An unfavorable factor is the continuing depletion of existing reserves, especially in the lower 48, where few major discoveries have been made in recent decades. Quantifying the depletion phenomenon is unfortunately problematic. In the United

---

[13] See American Petroleum Institute, *Basic Petroleum Data Book, 1975*, Section 2, Table 2.

[14] National Academy of Sciences, *Mineral Resources and the Environment* (Washington, D.C., 1975), p. 9. See also Edward Teller, *Energy: A Plan for Action* (New York: Commission on Critical Choices for Americans, 1975), esp. pp. 50-54.

[15] The National Academy of Sciences study cited in note 14, for instance, also states that "a large increase in annual production from conventional sources is extremely unlikely." The term "conventional sources" apparently excludes additional recovery from known reservoirs.

States, annual production now exceeds 10 percent of "proved reserves," but it does not follow that the depletion rate is 10 percent. The authorities who estimate proved reserves habitually make an annual correction known as "revisions and extensions," which in recent years has amounted to about two-thirds of annual production. This practice appears to reflect the very conservative basis on which proved reserves are initially estimated. Moreover, some discoveries would be made in the United States and Canada even if prices had remained at their pre-1973 levels. On balance, an annual decline rate of 5 percent would seem to be appropriate for North American production, and it has also been used for other areas.

The third factor affecting future output in North America is discoveries already made but not yet in production. The outstanding example, of course, is Alaska. Although the Prudhoe Bay field was discovered in 1967, the first oil is not expected until 1977. However, neither the delay nor the ultimate go-ahead for the pipeline had much to do with the price of crude. For projection purposes the crude from Alaska and a few similar but less important areas such as offshore California and Elk Hills should be treated as a "fixed supply" in years after 1977, subject to neither price elasticity nor decline.

Fixed supplies are even more important in Europe, where otherwise 1972 production was so small that any discussion of price elasticity and depletion rates would be superfluous. North Sea oil, however, is clearly going to be important in the world market, although the size of the reserves is still uncertain. Since past discoveries are still being developed and new discoveries continue to be made, optimistic projections suggest that North Sea production could take care of the bulk of European consumption by the middle 1980s, but the many technical and political uncertainties make it advisable to postulate a more modest fixed supply, which would leave Europe a large net importer at almost any price.

## Implementation of the World Petroleum Model

So far this paper has given a verbal description of a quantitative model of the world petroleum market, or more precisely of the simplest version of this model.[16] Keeping track of the many ramifications of developments in the world petroleum market calls for an explicit model that permits policy simulations on a computer. The

---

[16] For details see Kennedy, "Economic Model," and his article in D. W. Jorgenson, *Econometric Studies of U.S. Energy Policy* (New York: North Holland-American Elsevier, 1976).

version used in this paper divides the non-Communist world into six regions (the United States, Canada, Latin America, Europe, the Middle East together with Africa, and the rest of Asia together with Australia). It recognizes four products[17] (gasoline, kerosene, distillate fuel oil, and residual fuel oil) but does not explicitly distinguish among different types of crude, although such a distinction is implicit in the refining coefficients.

The model takes 1972 data on production, trade, consumption, prices, and taxes as a starting point, and then calculates the consequences of changing any of the multitudinous parameters that describe the world petroleum market. The version used here is designed primarily for policy simulations referring to any year between 1980 and 1985. The initial year is probably far enough away to justify the use of long-run elasticities of supply and demand and the neglect of inventory fluctuations; the final year is probably not so far away that any of the technological assumptions (for instance, a continuing minor role for nuclear power) becomes invalid.

Each run of the model produces a solution describing production, trade, consumption, and the prices of crude and products in each of the six regions. The principal number to be specified is the export tax on crude from the Persian Gulf; in addition, any number of other parameters, such as elasticities or taxes of various kinds, can be reset. The two sets of values (optimistic and pessimistic) for the various elasticities that underlie this paper are given in Table 4. The model uses 1972 dollars throughout so that all prices are relative to an unspecified general price level.[18] The income elasticities in Table 1 are used in conjunction with regional growth rates (also in constant dollars), which for this paper were set at 2.5 percent for the United States, 3.5 percent for Canada and Europe, 5.5 percent for Latin America, 6 percent for the Middle East and Africa, and 4.5 percent for Asia-Australia, all starting from 1972. Of the many other parameters in the model, the only ones that require mention here are the fixed supplies (from oil fields previously discovered but not being exploited in 1972): they were assumed to be 1.8 million barrels per day in the United States (on account of Alaska) and 3.2 mmb/d in Europe (on account of the North Sea), both in 1980. In addition, fixed supplies of 0.5 mmb/d each were assumed for Latin America (to represent Mexican exports) and for Asia-Australia (to take account of exports from mainland China).

---

[17] Other products are considered a deduction from refining costs.
[18] A 1972 dollar was equal to about $1.25 in 1975 and may be equivalent to about $1.60 in 1980, assuming an annual inflation rate of 5 percent.

## Table 4
## LONG-TERM ELASTICITIES: OPTIMISTIC AND PESSIMISTIC ASSUMPTIONS

| Region | Crude Supply | Gasoline | | Kerosene | | Distillate | | Residual | |
|---|---|---|---|---|---|---|---|---|---|
| | | Price | Income | Price | Income | Price | Income | Price | Income |
| United States | | | | | | | | | |
| Optimistic | 0.5 | −1.0 | 1.0 | −0.5 | 1.5 | −0.5 | 0.6 | −1.0 | 1.5 |
| Pessimistic | 0.15 | −0.25 | 1.0 | −0.3 | 1.5 | −0.2 | 0.6 | −0.75 | 1.5 |
| Canada | | | | | | | | | |
| Optimistic | 0.5 | −1.0 | 1.0 | −0.5 | 1.5 | −0.5 | 0.6 | −1.0 | 1.5 |
| Pessimistic | 0.15 | −0.25 | 1.0 | −0.3 | 1.5 | −0.2 | 0.6 | −0.75 | 1.5 |
| Latin America | | | | | | | | | |
| Optimistic | 0.1 | −1.0 | 1.5 | −0.5 | 0.5 | −0.5 | 1.0 | −1.0 | 1.5 |
| Pessimistic | 0.05 | −0.6 | 1.5 | −0.4 | 0.5 | −0.6 | 1.0 | −0.8 | 1.5 |
| Europe | | | | | | | | | |
| Optimistic | 0.1 | −1.0 | 1.3 | −0.5 | 1.0 | −0.5 | 1.2 | −1.0 | 1.5 |
| Pessimistic | 0.05 | −0.5 | 1.3 | −0.3 | 1.0 | −0.6 | 1.2 | −0.8 | 1.5 |
| Middle East and Africa | | | | | | | | | |
| Optimistic | 50.0 | −1.0 | 1.6 | −0.5 | 1.0 | −0.5 | 0.8 | −1.0 | 1.5 |
| Pessimistic | 0.1 | −0.75 | 1.3 | −0.4 | 1.0 | −0.6 | 1.0 | −0.8 | 1.5 |

## Simulation Results with "Optimistic" Elasticities

In presenting results, we shall consider in particular the consequences of alternative assumptions for the centrally important price of crude in the Persian Gulf. The crude price in that area (which includes Africa in this version of the model) has two components, the cost of producing the oil and a notional export tax. The first depends on the volume of output and is determined within the model; for large volumes it is about 30 cents per barrel but it drops to 11 cents for small volumes when only the lowest-cost wells are supposed to be in operation. For each of several rates of this export tax in 1980 Tables 5A and 5B show certain important prices, production and trade volumes, and revenues. The more complete description of the world petroleum market produced by the model would take up far too much space.

Under the "optimistic" elasticity assumptions underlying Table 5A the trade patterns in the world market would clearly be very sensitive to the export tax rate in the Middle East and Africa (ME&A).[19] At a tax rate of $1.25 per barrel, the U.S. price of crude would be $3.05, roughly what it was around 1970. U.S. production of crude, condensate, and natural gas liquids would then be only 8.9 mmb/d, including the fixed supply from Alaska of 1.8 mmb/d. Consumption of the four main products would be 16.6 mmb/d, 26 percent more than the actual consumption of 1972. According to the submodel of the refining sector, and allowing for some product imports, this demand would require 20.5 mmb/d of crude and the like, of which 11.6 mmb/d, or 57 percent, would have to come from abroad. European product consumption of 15.6 mmb/d would require crude imports of 16.0 mmb/d. Latin America would supply part of these imports, but the ME&A region would supply not only the remaining U.S. and European demand but also the large Asian demand, primarily from Japan.[20] Export tax revenues in the ME&A region would be

---

[19] An earlier version of this table appeared in Edward J. Mitchell, ed., *Dialogue on World Oil* (Washington, D.C.: American Enterprise Institute, 1974). It was also based on the "optimistic" elasticities of Table 4, but assumed somewhat larger fixed supplies in the United States and Europe, a lower decline rate, and higher growth rates than Table 5A.

[20] Actually, the model understates Asian imports because the simple version used here assumes that all Asian production (chiefly from Indonesia) is consumed within the region. Because of its desirable characteristics, some Indonesian crude is shipped to the United States, and therefore has to be replaced by Middle Eastern crude.

## Table 5A
## IMPLICATIONS OF ALTERNATIVE EXPORT TAX RATES ON CRUDE OIL IN THE MIDDLE EAST, 1980—OPTIMISTIC ELASTICITY ASSUMPTIONS, NO NEW IMPORT DUTIES

| Export Tax Rate (1) | Crude Price[a] | | | Crude Production | | Product Consumption[b] | | Net Crude Exports | | | Tax Revenue, ME&A[c] (12) |
|---|---|---|---|---|---|---|---|---|---|---|---|
| | ME&A[c] (2) | U.S. (3) | Europe (4) | ME&A[c] (5) | U.S. (6) | U.S. (7) | Europe (8) | ME&A[c] (9) | U.S. (10) | Europe (11) | |
| 1.25 | 1.55 | 3.05 | 2.55 | 43.2 | 8.9 | 16.6 | 15.6 | 38.7 | −11.6 | −16.0 | 17.7 |
| 2.50 | 2.81 | 4.31 | 3.81 | 33.6 | 10.2 | 14.8 | 13.6 | 29.8 | − 8.0 | −13.5 | 27.2 |
| 3.75 | 4.06 | 5.56 | 5.06 | 26.8 | 11.3 | 13.4 | 12.4 | 23.5 | − 5.2 | −12.0 | 32.1 |
| 5.00 | 5.29 | 6.79 | 6.29 | 21.8 | 12.3 | 12.4 | 11.5 | 18.8 | − 2.9 | −11.5 | 34.3 |
| 6.25 | 6.52 | 8.02 | 7.52 | 17.8 | 13.2 | 11.5 | 10.9 | 15.0 | − 0.9 | −10.2 | 34.3 |
| 7.50 | 7.75 | 8.54 | 8.75 | 15.4 | 13.6 | 11.1 | 10.4 | 12.9 | 0 | − 9.6 | 35.3 |
| 8.75 | 8.97 | 9.38 | 9.97 | 12.8 | 14.2 | 10.7 | 9.9 | 10.4 | 0 | − 8.4 | 33.2 |
| 10.00 | 10.18 | 10.53 | 11.18 | 10.1 | 14.9 | 10.1 | 9.5 | 7.9 | 1.4 | − 8.0 | 28.6 |

Units: Columns 1-4: 1972 dollars per barrel; columns 5-11: millions of barrels per day; column 12: billions of 1972 dollars.
[a] Crude includes condensate and natural gas liquids.
[b] Products consist only of gasoline, kerosene, distillate, and residual fuel oil.
[c] Middle East and Africa.

$17.7 billion if that region were able to meet the total demand of 43.2 mmb/d on its crude.[21]

Contrast this with the trade pattern at an ME&A tax rate of $10 per barrel. Under the "optimistic" elasticity assumptions, product consumption in the United States and Europe, and for that matter everywhere else, would be much lower than in the $1.25 case because of higher prices; indeed, it would be much lower than in 1972. Stimulated by a domestic crude price in excess of $10 (1972 value), crude production in the United States would be up to nearly 15 mmb/d, enough to leave a surplus of 1.4 mmb/d available for export. While Europe would still be a considerable importer, at a rate of 8 mmb/d, its needs would be met largely by exporters outside the ME&A region. Most of the relatively small ME&A exports of 7.9 mmb/d would go to Asia. Export tax revenues in the ME&A region would be $28.6 billion and total crude output there would be only 10.1 mmb/d.

It should also be noted that with a $10 ME&A tax rate, the price of U.S. crude is only 35 cents higher than the price of ME&A crude, while in the previous case (and for all tax rates up to $6.25) it is $1.50 higher. The latter differential corresponds to freight and the present U.S. import duty of 63 cents per barrel (assumed to continue), but when the United States ceases to import from the Middle East and Africa, that relation becomes irrelevant. Indeed, at tax rates still higher than those in Table 5A, the U.S. crude price would fall below the ME&A counterpart with which it would have to compete in export markets.

Evidently, under the assumptions made so far, a tax rate of $10 would price ME&A crude out of more markets than is consistent with revenue maximization. The table suggests that ME&A revenue would be maximized at a tax rate of about $7.50 per barrel, and this is confirmed by more detailed calculations.[22] The maximum obtainable revenue in 1980 would be about $35 billion if the assumptions behind Table 5A were realized. The corresponding ME&A crude output would be only 15.4 mmb/d, compared with 27.1 mmb/d in 1973 and 24.6

---

[21] This figure includes the crude input into local product consumption. Contrary to the statement by B. A. Kalymon, "Economic Incentives in OPEC Oil Pricing Policy," *Journal of Development Economics*, vol. 2 (December 1975), p. 339, the model does take the domestic consumption of oil exporters into account.

[22] The demand curve for ME&A crude obtained by combining columns (1) and (9) of Table 5A is actually more complicated than inspection of that summary table might suggest. The curve has some local maxima at points where trade patterns change qualitatively.

mmb/d in 1975. Revenue maximization would therefore imply a further substantial cutback in output, whose consequences for OPEC will be discussed below.

## Simulation Results with "Pessimistic" Elasticity Assumptions

Crucial to these policy questions is the sensitivity of the results in Table 5A to changes in the more important assumptions. In particular, the "optimistic" price elasticities in Table 4 are open to question in the light of recent experience. This uncertainty affects especially the supply elasticities which, for the higher tax rates in Table 5A, imply 1980 levels of U.S. crude output that are widely believed to be higher than is physically possible. Although the demand projections are subject to less doubt, alternative assumptions may help in assessing the range of error in the projections.

Table 5B gives solutions for the World Petroleum Model with the "pessimistic" elasticity assumptions of Table 4; all other parameters are the same as in Table 5A. The most serious objection to the previous projections has now been largely removed: even at a price for U.S. crude as high as $24 (1972 value), domestic production would be only 11.8 mmb/d, about the same as actual production in 1975 if natural gas liquids and Alaskan oil are counted. Even at this high price the United States remains a net importer. Product consumption in the United States is also closer to the 1975 level: at an ME&A tax rate of $7.50 it would be 13.3 mmb/d, compared with 11.1 mmb/d under the assumptions of Table 5A. This similarity to 1975 is no guarantee of accuracy in 1980, but it may be reassuring to those who do not believe in price elasticities anyway.

The trade patterns shown in Table 5B, however, certainly are not reassuring in any other respect, except perhaps to OPEC. The most striking feature is the final column, 12, according to which the largest ME&A tax revenue occurs at a tax rate of slightly less than $20 per barrel (in 1972 dollars, corresponding to at least $30 in 1980 dollars). Unlike the case in Table 5A, consumers can take no consolation that the maximum is rather flat so that the revenue would not be much less at a tax rate of $15 per barrel than at $20; even a $15 tax rate would mean a doubling of present prices in current dollars.

However, the maximum revenue under pessimistic assumptions occurs at an ME&A output of about 13 mmb/d, even less than the revenue-maximizing output in Table 5A. But is it likely that OPEC could cut output so drastically?

## Table 5B

PROJECTED IMPLICATIONS OF ALTERNATIVE EXPORT TAX RATES IN THE MIDDLE EAST, 1980—PESSIMISTIC ELASTICITY ASSUMPTIONS, NO NEW IMPORT DUTIES

| Export Tax Rate (1) | Crude Price | | | Crude Production | | | Product Consumption | | Net Crude Exports | | | Tax Revenue, ME&A (12) |
|---|---|---|---|---|---|---|---|---|---|---|---|---|
| | ME&A (2) | U.S. (3) | Europe (4) | ME&A (5) | U.S. (6) | U.S. (7) | Europe (8) | ME&A (9) | U.S. (10) | Europe (11) | | |
| 2.50  | 2.82  | 4.32  | 3.82  | 36.5 | 9.5  | 15.4 | 14.0 | 32.6 | −9.4 | −14.0 | 29.7 |
| 5.00  | 5.33  | 6.83  | 6.33  | 28.5 | 10.0 | 14.1 | 12.1 | 25.4 | −7.3 | −11.7 | 46.3 |
| 7.50  | 7.83  | 9.33  | 8.83  | 23.7 | 10.4 | 13.3 | 11.0 | 21.0 | −5.9 | −10.3 | 57.4 |
| 10.00 | 10.31 | 11.81 | 11.31 | 20.3 | 10.8 | 12.6 | 10.2 | 17.8 | −4.8 | −9.4  | 65.1 |
| 12.50 | 12.78 | 14.28 | 13.78 | 17.6 | 11.0 | 12.2 | 9.6  | 15.4 | −3.9 | −8.6  | 70.3 |
| 15.00 | 15.24 | 16.74 | 16.24 | 15.4 | 11.2 | 11.8 | 9.1  | 13.4 | −3.2 | −8.0  | 73.4 |
| 17.50 | 17.70 | 19.20 | 18.70 | 13.7 | 11.4 | 11.4 | 8.7  | 11.8 | −2.6 | −7.4  | 75.1 |
| 20.00 | 20.16 | 21.66 | 21.16 | 12.1 | 11.6 | 11.1 | 8.3  | 10.3 | −2.0 | −7.0  | 75.3 |
| 22.50 | 22.61 | 24.11 | 23.61 | 10.8 | 11.8 | 10.9 | 8.0  | 9.1  | −1.6 | −6.6  | 74.3 |

**Note:** For notes and units see Table 5A.

## Constraints on OPEC

From modest beginnings in the early 1960s, the Organization of Petroleum Exporting Countries gradually built up its power and expertise until by late 1973 it was able to enforce its pricing policy without encountering significant resistance from consuming countries or from the international petroleum companies. Its thirteen members[23] have so far managed to maintain a common front despite occasional bickering, an achievement that is all the more remarkable in light of the ancient animosities dividing some of them. After the quadrupling of prices at the start of 1974, another increase was agreed upon in the late summer of 1975; although it apparently has not been implemented fully, the overall price level has held firm. Predictions (including my own) that OPEC would have to reduce its price because of declining demand and increasing outside supplies have not been realized. To be sure, the cartel has had considerable help from consuming countries, not least from the United States where a long debate over energy policy has so far produced little of value; the bill finally enacted in December 1975 is more concerned with punishing the major oil companies than with making the United States less vulnerable to monopolistic exploitation by OPEC. Environmentalist opposition to offshore drilling—the best prospect for the longer run in fossil fuels—and to nuclear power remains strong. Perhaps the most positive achievements in this field were the opening of the Naval Petroleum Reserves to more intensive exploitation and exploration and the start of a stockpiling program.

With OPEC so firmly in the saddle, what can we expect for the coming years? In particular, will OPEC be able to adopt the revenue-maximizing tax rates suggested by Tables 5A and 5B?

Answers to these questions rest on several developments. First, world demand for oil has already fallen significantly (see Table 1), and the decline appears to be due largely to high prices and not merely to the recession. Although non-OPEC production, especially in North America, has also been relatively weak, most OPEC members have in fact had to reduce their output, as shown in Table 6. For completeness this table also shows non-OPEC production.

What is remarkable about the OPEC figures in Table 6 is their diversity. Roughly speaking, there are three groups of members. The first and largest includes the two leading countries, Saudi Arabia and Iran, and also Nigeria and Indonesia; in 1975 the output of these countries had been reduced from its 1973 level by about the same

---

[23] Counting the United Arab Emirates as one.

## Table 6
### WORLD OUTPUT OF CRUDE OIL[a]
(in millions of barrels per day)

| Area | 1973 | 1974 | 1975 | 1975 as Percent of 1973[b] |
|---|---|---|---|---|
| WORLD | 55.7 | 55.9 | 53.2 | 95 |
| NON-COMMUNIST AREAS | 45.8 | 45.1 | 41.5 | 91 |
| OPEC, Total | 30.9 | 30.5 | 27.2 | 88 |
| Saudi Arabia | 7.6 | 8.5 | 7.1 | 93 |
| Iran | 5.9 | 6.0 | 5.4 | 91 |
| Venezuela | 3.4 | 3.0 | 2.3 | 70 |
| Iraq | 2.0 | 1.8 | 2.3 | 115 |
| Kuwait | 3.0 | 2.5 | 2.1 | 69 |
| Nigeria | 2.1 | 2.3 | 1.8 | 87 |
| United Arab Emirates | 1.5 | 1.7 | 1.7 | 131 |
| Libya | 2.2 | 1.5 | 1.5 | 69 |
| Indonesia | 1.3 | 1.4 | 1.3 | 98 |
| Other OPEC | 1.9 | 1.8 | 1.7 | 90 |
| OECD, Total | 11.8 | 11.3 | 10.9 | 92 |
| United States | 9.2 | 8.8 | 8.4 | 91 |
| Canada | 1.8 | 1.7 | 1.5 | 82 |
| Western Europe | 0.4 | 0.4 | 0.6 | 150 |
| Australia | 0.4 | 0.4 | 0.4 | 88 |
| Other non-Communist | 3.1 | 3.3 | 3.4 | 110 |
| COMMUNIST AREAS | 9.9 | 10.7 | 11.6 | 118 |
| U.S.S.R. | 8.4 | 9.0 | 9.6 | 114 |
| China | 1.1 | 1.3 | 1.6 | 149 |
| Other Communist | 0.4 | 0.4 | 0.4 | 104 |

[a] Does not include natural gas liquids.
[b] Calculated from unrounded figures.
**Source:** Same as Table 1.

percentage as the fall in world production, so that market shares were maintained. The next largest group consists of Venezuela, Kuwait, and Libya, all of which cut output by some 30 percent below 1973. For the remaining group, Iraq and the United Arab Emirates (principally Abu Dhabi), crude production actually went up considerably. It is too early to say whether this grouping will endure but clearly OPEC has not been able or willing so far to allocate output among its members.

The share of OPEC in the world market as a whole fell from 55 percent in 1973 to 51 percent in 1975, with the U.S.S.R. and China the chief gainers.

OPEC would therefore find it difficult to cut its output as drastically as would be needed to maximize 1980 revenue. The current dispute about the comparatively minor problems of price differentials among crudes of different origins suggests that OPEC is not ready to take on the much more delicate problem of prorationing. While Saudi Arabia, Iran, and Venezuela are large enough to consider the effect of their output decisions on the cartel as a whole, this is not true of several other members, some of which (especially Iraq and Libya) have a history of independent behavior. On the other hand, all members are well aware of their potential losses if the cartel were to break down, so there are limits on independence. Perhaps an uneasy equilibrium at a total output level not much below that of 1975 is the most likely outcome for the next several years.

For this reason, the revenue-maximizing tax rates calculated from Tables 5A and 5B are probably academic.[24] The only way in which a cartel could come close to these tax rates is by allowing side payments, by which reluctant members would be induced to participate in a prorationing scheme. In practice, Saudi Arabia and Kuwait would have to bribe Iraq and Abu Dhabi to reduce their output drastically; other leading OPEC members such as Iran are already importing so much that they would be hard pressed to contribute. While the introduction of side payments is not inconceivable, the practice would encounter serious political obstacles such as the philosophical differences between conservative Saudi Arabia and radical Iraq.

During the next two or three years the oil market is likely to come under increasing pressure from new supplies, particularly from Alaska and the North Sea. Active exploration in many parts of the world may open up still more new oil provinces. Although a recovery from the worldwide recession is now in progress, this by itself is not likely to raise demand sufficiently to offset the continuing adjustment

---

[24] This remark applies even more strongly to the optimization of OPEC revenue over time, which is the principal subject of the research surveyed by Fischer, Gately, and Kyle in "Prospects for OPEC." While this approach is theoretically superior to the single-period optimization considered above, it is open to serious practical objections, in particular the need to estimate oil prices in the more distant future. The naive view that petroleum must become ever more valuable because there is only a finite amount in the ground ignores two plausible possibilities: (1) major oil discoveries will be made, and (2) other sources of energy (such as nuclear) will become cheap enough to displace oil from important markets.

to higher energy prices. OPEC will therefore have its hands full maintaining the 1975 level of output, even under the "pessimistic" assumptions made earlier.

An additional problem faced by OPEC is increasing competition from non-OPEC exporters. According to Table 6, these are still of minor importance, but the incentive to grow is strong. Mexico, Oman, and Malaysia-Brunei are the most likely to make inroads into OPEC markets; the first two are already producing well above their 1973 levels. Egypt and Syria are also coming on strongly, but they are heavily dependent on other Arab countries and may join OPEC if their production continues to grow. Among the developed countries, Canada and Norway also could be a threat to OPEC, but so far have behaved like informal members of the cartel: Canada has discouraged domestic production and exploration by export duties, while Norway has restrained the development of the large fields discovered in its part of the North Sea. While unlikely to become exporters, such importing countries as India, Brazil, and Spain have already made significant progress toward self-sufficiency through exploration.

And then there is the danger to OPEC of substitution from other fuels, especially coal. Coal production in the United States has not increased much (in large part because many electric power stations are no longer equipped to burn coal), but a large expansion is possible if and when environmental objections can be resolved. The price of coal is now high enough to permit exploitation of the vast deposits of low-grade (but also low-sulphur) coal in the Mountain States, yet not so high as to make coal unattractive for electric power generation. The prospects for synthetic oil from oil shale and tar sands are still cloudy; the early enthusiasm for the first has not survived better cost calculations, while the outlook for the second has been adversely affected by Canadian government policies. Nuclear power, however, now appears to have a cost advantage over fossil fuels in many parts of the United States, and also in Europe and Japan; here, too, the main problems are environmental, while a marked slowdown in the growth of electricity consumption has made the utility companies reluctant to invest in new facilities of any kind.

None of this means that the era of high oil prices, which has only just begun, is likely to end soon. The purpose of enumerating the various constraints on OPEC, most of which do not appear explicitly in the World Petroleum Model, is to qualify the possible inference from the "pessimistic" projections that another large price increase is in the cards before 1980. While such an increase certainly cannot be ruled out, its probability can be reduced by appropriate policies on the part of consuming countries.

## Possible Countermeasures by Oil-Importing Nations

What can the consuming countries do to weaken the market power of OPEC, and thus to obtain lower prices? One set of responses, of course, has already been taken into account in this analysis: consuming less and producing more. Although exhortation may reinforce them, these reactions are mostly automatic and induced by higher prices. Whether they are sufficient depends on the supply and demand elasticities, and the "optimistic" and "pessimistic" assumptions analyzed above give some indication of their effectiveness, though not necessarily of the whole range of possible outcomes. At least until 1980, not much relief can be expected on this score even if the elasticities are higher than is generally believed.

More aggressive countermeasures could be considered.[25] I neglect strictly political approaches since they would require more cohesion among the importers than can be expected at this time. In any case, the potential of such measures is hard to analyze by the methods of this paper.

In the medium term considered here the principal positive countermeasures take the form of taxes, tariffs, or subsidies, or some combination of the three. Excise taxes on products would have a depressing effect on consumption, while subsidies could be used to encourage domestic output. In view of congressional attachment to price controls and popular dislike of the oil industry, neither of these measures has much chance of adoption in the foreseeable future. A tariff on imported oil would have much the same economic effect on production and consumption as an excise tax combined with an output subsidy, but it has the political advantage of not involving outright payments to producers and of being levied only on imports.[26] Moreover, a tariff could conceivably be adopted internationally among all the consuming countries, which would greatly strengthen its impact on OPEC.

Without implying that it would be very palatable or easily arranged, what could be accomplished by a uniform tariff on crude and products levied by the members of the International Energy Agency

---

[25] Such measures include the military intervention hinted at by Secretary Kissinger as a last resort in case of "strangulation."
[26] Admittedly, the $3 import duty announced by President Ford in January 1975 (only $2 of which was actually imposed, until the duty was withdrawn altogether at the end of 1975) also encountered strong opposition. However, a major objection was that its economic effects would be largely nullified by price controls on domestic output. The analysis here (as elsewhere in this paper) assumes that price controls will be phased out well before 1980.

## Table 7
### PROJECTED IMPLICATIONS OF ALTERNATIVE COMBINATIONS OF EXPORT AND IMPORT TAXES, 1980, TWO SETS OF ELASTICITY ASSUMPTIONS

| ME&A Export Tax Rate ($/bl) | | IEA Import Tax Rate ($/bl) | | | | | |
|---|---|---|---|---|---|---|---|
| | | 0 | | 2.50 | | 5.00 | |
| | | ME&A | IEA | ME&A | IEA | ME&A | IEA |
| | | Optimistic elasticities | | | | | |
| 3.75 | mmb/d | 26.8 | 23.5 | 19.1 | 15.7 | 15.9 | 12.5 |
| | $ bln | 32.1 | 0 | 21.6 | 14.4 | 17.1 | 22.8 |
| 5.00 | mmb/d | 21.7 | 19.1 | 16.5 | 13.7 | 14.6 | 11.6 |
| | $ bln | 29.4 | 0 | 24.7 | 12.5 | 21.0 | 21.2 |
| 6.25 | mmb/d | 17.8 | 15.6 | 14.9 | 12.6 | 13.1 | 10.8 |
| | $ bln | 34.3 | 0 | 27.6 | 11.5 | 23.4 | 19.7 |
| 7.50 | mmb/d | 15.4 | 13.6 | 12.7 | 10.9 | 11.1 | 9.3 |
| | $ bln | 35.3 | 0 | 27.7 | 10.0 | 23.5 | 17.1 |
| 8.75 | mmb/d | 12.8 | 11.4 | 10.8 | 9.4 | 8.9 | 7.5 |
| | $ bln | 33.2 | 0 | 26.8 | 8.6 | 20.6 | 13.6 |
| | | Pessimistic elasticities | | | | | |
| 5.00 | mmb/d | 28.5 | 25.4 | 24.6 | 21.4 | 21.7 | 18.5 |
| | $ bln | 46.3 | 0 | 39.1 | 19.5 | 33.7 | 33.7 |
| 7.50 | mmb/d | 23.7 | 21.4 | 20.9 | 18.5 | 18.6 | 16.3 |
| | $ bln | 57.4 | 0 | 49.6 | 16.9 | 43.5 | 29.7 |
| 10.00 | mmb/d | 20.3 | 18.6 | 18.1 | 16.3 | 16.2 | 14.5 |
| | $ bln | 65.1 | 0 | 57.0 | 14.9 | 50.3 | 26.4 |
| 12.50 | mmb/d | 17.6 | 16.4 | 15.8 | 14.5 | 14.3 | 13.0 |
| | $ bln | 70.3 | 0 | 61.9 | 13.2 | 55.0 | 23.7 |
| 15.00 | mmb/d | 15.5 | 14.6 | 14.0 | 13.0 | 12.7 | 11.8 |
| | $ bln | 73.4 | 0 | 65.2 | 11.9 | 58.1 | 21.5 |

**Notes:** Of the four figures for each tax combination, the upper left gives total crude output in the Middle East and Africa, the upper right gives total imports of crude and products in North America, Europe and Asia, the lower left gives export tax revenue in the Middle East and Africa, and the lower right gives import tax revenue in the consuming areas. All money figures are in 1972 dollars.

(Europe, North America, and Japan)? This tariff is assumed to be an increment to any existing national tariffs and to be nondiscriminatory; thus, it would apply to any exports from North America to the other consuming areas. Table 7 provides projections for 1980 on "optimistic" and "pessimistic" elasticity assumptions. The first column, corresponding to a zero import tax, overlaps with Tables 5A and 5B. All money figures, as before, are in 1972 dollars.

Take, for instance, an export tax rate of $7.50, roughly the present level after translation to 1976 dollars. Under pessimistic elasticity assumptions, and in the absence of an import tax, this rate would imply a crude output of 23.7 mmb/d in the Middle East and Africa, again close to the present level. Total imports of the International Energy Agency countries (IEA)[27] would be 21.4 mmb/d and ME&A export tax revenues would be $57.4 billion.

Suppose now that the IEA imposed an import tax of $2.50 per barrel. This would reduce IEA imports by some 3 mmb/d and ME&A output by almost the same amount; there would also be a minor reduction in output in Latin America not shown in the table. Tax revenues would be $49.6 billion in the ME&A region and $16.9 billion in the IEA region. Similarly, an IEA import tax of $5 per barrel would reduce ME&A output to 18.6 mmb/d and ME&A revenues to $43.5 billion, while giving IEA a revenue of $29.7 billion.

At an ME&A tax of $7.50 and an IEA tax of $2.50, the trade volumes would be approximately the same as at an ME&A tax of $10 and no IEA tax; the principal difference, apart from some secondary effects on Latin America, would be in the cost of IEA imports and in the respective tax revenues. The IEA import bill, not shown in the table, is closely correlated with the ME&A export tax revenue, differing from it only by the production cost in the exporting countries, the value of imports from Latin America, and the freight charges.[28] The pattern of trade depends primarily on the *total* tax per barrel, no matter how it is divided between exports and imports. Conversely, for a given volume of trade, every tax dollar collected by IEA is a dollar saved on its import bill, and this dollar is paid for entirely by the producers' cartel.

The significance of this observation for oil policy derives from the earlier observation that OPEC is not likely to set its export tax

---

[27] This is not strictly correct because the Asian region, in particular, includes many nonmembers of IEA and one OPEC member (Indonesia). Also France, which is not a member of IEA, has been included. The simple version of the World Petroleum Model used here does not permit a more refined breakdown.

[28] With a $7.50 export tax and a $2.50 import tax, for instance, the total value of imports (excluding duty) in the importing areas would be $59.8 billion, $17.1 billion for the U.S., $0.8 billion for Canada, $29.8 billion for Europe, and $12.1 billion for Asia and Australia. With a $10 export tax and no import tax, the value of IEA imports would be $76.9 billion (all in dollars of 1972). The difference of $17.1 billion between the two import bills is almost exactly equal to the $16.9 billion collected by IEA from a $2.50 tax. This approximate relationship holds throughout Table 7.

rate at the revenue-maximizing level because its output would then be so low as to make the cartel unviable.[29] Even at the present OPEC output, some cartel members are unhappy, and any further reduction would create acute dissension. The detailed analysis necessary to locate the probable breaking point cannot be attempted here.[30]

Assume, for illustrative purposes, that the lowest ME&A daily output consistent with reasonable OPEC cohesion in 1980 is about 20 million barrels (say, 5 million barrels for Saudi Arabia; 4½ million for Iran, 2 million for Iraq; 1½ million each for Kuwait, the United Arab Emirates, and Nigeria; 1 million each for Libya and Algeria; and 2 million for all other ME&A producers, most of whom do not belong to OPEC). According to Table 7, assuming pessimistic elasticities, this level of output could be achieved with an ME&A export tax of about $10 per barrel if there were no IEA tax. Total IEA imports would then be 18.6 million barrels per day, virtually the same as with a $7.50 export tax and a $2.50 import tax, or with a $5 tax on both exports and imports. However, ME&A tax revenues in these three cases would be $65.1 billion, $49.6 billion, and $33.7 billion respectively, while IEA tax revenues would be zero, $16.9 billion, and $33.7 billion. The lower ME&A tax revenue due to an IEA tariff would aggravate the strains inside the cartel (in addition to the low level of output) since members with balance-of-payments problems would be more inclined to seek special deals with oil importers.

At the moment, the ME&A export tax is equivalent to about $7.50 per barrel. If 20 million barrels per day were indeed the minimum at which the cartel could survive, it would be tempted to raise its export tax rate by one-third in real terms between now and 1980.[31] The importing countries may be able to prevent this by imposing a common import tax of $2.50 per barrel (about $3.25 in current dollars). An even higher import tax would be a direct threat to OPEC stability.

---

[29] Note that the effect of an import tax is *not* to reduce the revenue-maximizing ME&A tax rate, as a look down the columns of Table 7 will confirm. Although the ME&A tax revenue is lower when consumers take countermeasures, its maximum (for a given IEA tax) still occurs at about the same point. This invariance is actually a consequence of the constant-elasticity form assumed here for the supply and demand functions, and would not necessarily apply for different functional forms.
[30] For a discussion of the circumstances of each OPEC member, see Philip K. Verleger, Jr., *World Petroleum Outlook* (New York: Data Resources, Inc., 1976), pp. 103-24.
[31] As Professor M. A. Adelman has pointed out, in the short run the easiest way for a cartel to restore internal harmony is to raise the price, even though doing so tends to worsen the underlying problem of allocating production in the longer run.

An IEA import tax, by transferring funds from OPEC back to oil importers, would improve the latter's balance of payments. To the extent that it prevents an increase in the cartel price, it would also benefit the developing countries that import oil but do not belong to IEA. However, this paper does not emphasize the purely financial aspects of higher oil prices because, contrary to the fears widely expressed in 1973–74, they are not of major importance.[32] The international financial system has so far proved quite capable of recycling the greatly increased flow of "petrodollars"; by and large, the surpluses accumulating in the hands of some petroleum exporters have been re-lent to those petroleum importers who were unable to expand their exports sufficiently. No doubt, the sudden quadrupling of crude prices at the end of 1973 contributed to the worldwide inflation of 1973–74 and the worldwide recession of 1974–75, but by now both of these are well on the way to improvement.

As to OPEC's reaction to an IEA import tax, there is not a great deal it could do. Most OPEC members are so dependent on export revenues that another embargo is unlikely; in any case the 1973–74 embargo was largely ineffectual in cutting off oil supplies to the embargoed countries.[33] The main adverse effect of an import tax might be a reinforcement of political cooperation within the cartel.

How does an import tax compare with an alternative countermeasure that has been extensively discussed in the International Energy Agency—a so-called floor price, usually put at $7 per barrel? The reasoning behind this proposal is the encouragement it would give to domestic oil output in certain IEA countries. While the high-cost production in the North Sea is usually mentioned as the prime example, the same considerations apply even more strongly to enhanced recovery from old fields in the United States. That a floor price would also give price protection to OPEC is not necessarily a disadvantage since it might encourage OPEC members to emancipate themselves from the cartel without fears of causing the world price to plummet.

---

[32] For a more extensive discussion, see Edward R. Fried and Charles L. Schultze, eds., *Higher Oil Prices and the World Economy* (Washington, D.C.: The Brookings Institution, 1975).

[33] For a discussion of the reshuffling of trade flows during the embargo, see Robert B. Stobaugh, "The Oil Companies in the Crisis," *Daedalus*, Fall 1975. The disruption of the U.S. petroleum market in the early months of 1974 was real enough, but resulted largely from official mismanagement and public hysteria; on this subject see Richard B. Mancke, *Performance of the Federal Energy Office*, National Energy Study 6 (Washington, D.C.: American Enterprise Institute, 1975).

The floor price is therefore an idea of considerable merit except if OPEC is expected to break up by itself, an outcome that no longer seems very probable. The main difficulty is quasi-political in that most IEA countries have little or no domestic production and are consequently reluctant to underwrite larger output in a few other developed countries. In this respect the IEA tariff considered here would seem more acceptable; it would also give some protection to IEA producers, but the other importers would gain from the revenue.

On the other hand, IEA is a new organization in which the interests of different members cannot always be easily reconciled. The question therefore arises, "Could the United States go it alone?" The World Petroleum Model provides detailed answers to this question under various assumptions; the main conclusion is that the United States is a big enough importer to exercise a significant effect on ME&A output by varying its import tax, but probably not big enough to threaten OPEC with collapse merely by its national measures.

Consider, for instance, the case of a $7.50 export tax under pessimistic elasticity assumptions. According to Table 7, ME&A output would then be 23.7 mmb/d in the absence of any tariff, and an IEA import tax of $5.00 per barrel would reduce ME&A output to 18.6 mmb/d, probably low enough to exert destructive strains within the cartel. An import tax levied exclusively by the United States would curtail ME&A production less than 10 percent to 21.5 mmb/d, a point at which the cartel is likely to survive. The U.S. oil import bill would fall from $20.8 billion to $13.1 billion per year, while the tariff would bring in some $7 billion (all in 1972 dollars).

These effects are by no means negligible, but a U.S. national tariff would also mean much higher energy costs here than in Europe and Japan, to the disadvantage of our nonoil trade. A common IEA tariff would therefore be preferable from the American point of view. The tariff could be even more effective if it exempted imports from IEA countries, a possibility not analyzed here.

**The Outlook for 1985**

So far, only projections for 1980 have been discussed, for the most part under pessimistic elasticity assumptions. To complete the analysis Table 8 considers 1985. It is similar in layout to Table 7 but is confined to optimistic elasticity assumptions, which are likely to be more valid in the longer run. All other assumptions (including those on decline rates, GNP growth rates, and fixed supplies) are maintained, so that Table 8 differs from the top half of Table 7 only in including

## Table 8
### PROJECTED IMPLICATION OF ALTERNATIVE COMBINATIONS OF IMPORT AND EXPORT TAX RATES, 1985, OPTIMISTIC ELASTICITY ASSUMPTIONS

| ME&A Export Tax Rate ($/bl) | | IEA Import Tax Rate ($/bl) | | | | | |
|---|---|---|---|---|---|---|---|
| | | 0 | | 2.50 | | 5.00 | |
| | | ME&A | IEA | ME&A | IEA | ME&A | IEA |
| 5.00 | mmb/d | 35.8 | 29.7 | 28.8 | 22.7 | 23.7 | 17.4 |
| | $ bln | 57.6 | 0 | 44.8 | 20.7 | 35.2 | 31.7 |
| 7.00 | mmb/d | 27.3 | 22.5 | 22.4 | 17.6 | 20.0 | 10.5 |
| | $ bln | 64.8 | 0 | 51.2 | 16.0 | 44.3 | 27.4 |
| 10.00 | mmb/d | 21.3 | 17.4 | 18.9 | 14.9 | 17.0 | 12.3 |
| | $ bln | 66.0 | 0 | 56.7 | 13.6 | 49.9 | 24.1 |
| 12.50 | mmb/d | 17.5 | 14.3 | 15.4 | 12.3 | 14.2 | 11.0 |
| | $ bln | 66.5 | 0 | 56.8 | 11.2 | 51.3 | 20.0 |
| 15.00 | mmb/d | 13.5 | 11.0 | 12.1 | 9.6 | 10.9 | 8.2 |
| | $ bln | 59.1 | 0 | 51.8 | 8.7 | 44.7 | 15.1 |

Notes: Same as for Table 7.

five more years of decline in production from existing fields and of growth in GNP.[34]

Table 8 suggests that in 1985, under optimistic elasticity assumptions, ME&A revenue would be maximized at an export tax rate of $12.50 per barrel; the equivalent of that figure in 1985 dollars would almost certainly be well over $20, depending on future inflation. This revenue-maximizing rate is again nearly invariant with respect to the rate of the IEA import tax (if any). At the present export tax rate of about $7.50 per barrel, ME&A output in 1985 would be somewhat greater than in 1975, though not by much. An IEA import tax of $5 per barrel could bring that figure down to 20 mmb/d, giving the ME&A region a tax revenue of $44.3 billion.

---

[34] Actually, the use of the same fixed supplies introduces an element of pessimism in Table 8. By 1985, several new major oil fields are likely to be in production, especially in the North Sea. On the basis of discoveries announced through the middle of 1976 it appears that North Sea production could easily reach 6 million barrels per day in 1985. No allowance has been made either for oil already discovered in the Canadian Arctic or for possible new sources in the Atlantic and elsewhere.

Needless to say, Table 8, dealing with a more remote future, is even more speculative than the other projections in this paper. Nevertheless, it reinforces the conclusion reached earlier that the consuming countries can hardly afford to ignore countermeasures against the cartel unless they are prepared to pay higher and higher prices in real terms.

## Conclusions

The analysis in this paper leads to the following conclusions.

First, projections of patterns of petroleum trade are sensitive to the assumptions made about the price elasticity of supply and demand. Two sets of elasticity assumptions were used to suggest a range of possible outcome.

Second, even under optimistic elasticity assumptions, the real price of oil is likely to be higher in 1980 and 1985 than it is now unless the consuming countries take appropriate countermeasures.

Third, OPEC is not likely to adopt an export tax rate that will maximize its revenues because at such a rate its output would be too small to preserve cohesion among the cartel members. Therefore, consumers' countermeasures should aim at reducing OPEC output to a level at which cartel cohesion becomes a problem.

Fourth, the most promising countermeasure is a common import tax levied by the member countries of the International Energy Agency. To prevent further OPEC price increases, this tax should probably be about $3.25 per barrel.

Finally, such a tax would have a favorable effect on the balance of payments of the oil importers because, at a given level of output, it would essentially be paid by producers.

## Postscript: The Longer Run

During the 1980s, the world energy situation will be increasingly influenced by the growth of nuclear power. At present, generation of electricity from uranium does not displace enough oil to have a significant effect on world crude prices, and therefore did not have to be recognized in the preceding medium-term analysis. In many parts of the world, however, electricity can now be produced from nuclear fuels at a lower cost than from oil and other fossil fuels, even if the higher capital requirements of nuclear stations are taken into account. This, of course, is why the electric power industry is interested in nuclear energy.

When the installation of nuclear stations is held down (as it is now) by long construction periods aggravated by regulatory delays, the demand for uranium does not strain the supply and its price can remain well below its oil equivalent. As more nuclear reactors are commissioned, the price of uranium will gradually be bid up to a level where either fossil fuels again become competitive, or the supply of uranium begins to exceed the demand. It is not known at the moment which of these two limits becomes operative first. Rough calculations suggest that, at an oil price of $10 per barrel, uranium oxide as an alternative source of electricity might be worth as much as $150 or $200 per pound, about four times the highest spot price reached in the last few years. In this price range, large quantities of low-grade uranium ores are believed to be available for commercial exploitation.

The importance of this tentative analysis for the world price of oil is that it can be turned around to derive the price of oil from the price of uranium, assuming sizable quantities of oil are in fact used for generating electricity anywhere in the world. Thus, if uranium oxide could be produced in large amounts at $100 per pound, then the price of oil would have to be between $5 and $7 per barrel; otherwise oil would have no market in the electric power industry. Although in the 1980s, oil may not be used for this purpose in the United States, it will continue to be so used elsewhere, and the equivalent argument is therefore relevant. In the longer run, in fact, uranium becomes a low-cost substitute for oil.[35]

---

[35] A similar argument applies to coal but there the new demand is not as large an addition to present consumption as it is in uranium. Because the present coal price is not greatly different from its oil equivalent, the incentive to build coal-fired power stations is not as strong as for nuclear stations.

Cover and book design: Pat Taylor

LIBRARY OF